aguaro Moon

A Desert Journal

and Illustrated

Joy Pratt-Serafini

Dawn Publications

The Sonoran Desert

My Planet Scout Journal
By Megan

6 June

Dear Journal,

Something exciting happened this summer—my family moved to a desert! To my surprise, it isn't the vast sandbox I was imagining at all. Arizona's Sonoran Desert is actually beautiful – but SO different. Instead of quiet conifers, there are spiny saguaros. Instead of lonely loons, there are racing roadrunners.

I started this new journal to record all the desert life I discover. I already miss my Planet Scout friends. I hope I make new friends here. Planet Scouts is a club for kids who like to study nature. We always keep a nature journal. I bring it with me whenever I go exploring. I used to take it to my secret thinking rock by the lake almost every week. I'll miss watching the waves and the clouds moving across the lake.

But there are plenty of new animals and plants for me to study here!

P.S. I really wish I had a friend to show me around and introduce me to all these strange species.

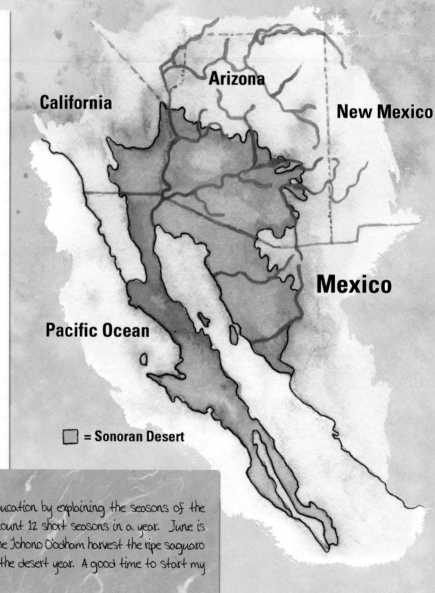

California

Arizona

New Mexico

Pacific Ocean

Mexico

☐ = Sonoran Desert

SEASON NOTE:
Mitchell began my desert education by explaining the seasons of the Sonoran Desert. His people count 12 short seasons in a year. June is called Saguaro Moon, when the Tohono O'odham harvest the ripe saguaro fruit. It is the beginning of the desert year. A good time to start my new journal!

What is a Desert?

Imagine a place where the air is dry, the sun is bright, and it doesn't rain very often. Plants and animals here have adapted to live without much water. You're in a desert!

Deserts cover approximately 8 million square miles – about one seventh of Earth's land surface. The Sahara Desert, which covers 3.5 million square miles of North Africa, is the largest desert in the world.

Bill's Big Biome Book   12

Megan   Mitchell

Meet the Sonoran Desert

By Sandy Banks, Tarantula Travel Times Staff

There are four major deserts in North America: the Chihuahuan, the Great Basin, the Mojave and the Sonoran. Together they cover roughly 500,000 square miles in the southwestern part of the United States and northwest Mexico.

The Sonoran is the only desert in North America that does not have cold winters. This makes it a tropical desert. It covers 100,000 square miles, and expanded to its present size only about 8-10,000 years ago, which makes it a medium-sized, but very young desert. The Sonoran Desert is also thought of as an arborous desert, which may seem like a contradiction, since "arborous" means trees. It is not that the Sonoran Desert is covered with forests, but there are tall cacti and many trees – it's more like a sparse woodland. Scientists have recorded more different types of animals and plants in the Sonoran than in any of the other North American deserts, making it a truly exciting place to explore.

7 June

This morning two neighbors, a boy and his grandmother, dropped by to welcome us to our new home. We invited them in for some lemonade. Mitchell and his Grandma Vavhia live about half a mile down the road. After visiting for a while, Grandma Vavhia asked if we would like to help harvest Saguaro fruit next week. (That's pronounced "sa-WAR-o", one of many Spanish words used around here.) That sounded interesting – and a great time to meet new friends! The adults kept talking for a long time, so Mitchell and I explored outside. He said that his family was part of the Tohono O'odham, a group of Native Americans who have lived in the Sonoran Desert for several hundred years. I just got here yesterday! Mitchell agreed to introduce me to the plants and animals of his homeland.

OCOTILLO

Fouquieria
splendens

DESERT
NIGHT
BLOOMING
CEREUS

Peniocereus greggii

ORGAN
PIPE
CACTUS

CARDÓN

Pachycereus
pringlei

Stenocereus
thurberi

SENITA

Lophocereus
schottii

CREEPING
DEVIL

Stenocereus
eruc

Saguaro Cactus
(Carnegiea gigantea)
HEIGHT: UP TO 50 FEET (15 M)

SAGUARO: A SONORAN SYMBOL

Saguaros (sa-WAR-os) are the distinguishing features of the Sonoran Desert. Their white, trumpet-shaped flowers are honored as the state flower of Arizona. Their life span is about 150 to 200 years, but they grow extremely slowly. Saguaros don't even start growing arms until they are about 50 to 70 years old. For this reason, the saguaros are protected by Arizona law. So don't expect to take one home as a desert souvenir. You'd have a tough time doing that anyway, since a mature saguaro can weigh up to 10 tons.

14 June

How strange saguaros look! In a way, they seemed more like people than plants. Some have lots of arms reaching out in different directions. I told Mitchell that, and he said the Tohono O'odham people think of them that way too.

The saguaro fruit harvest was my first desert adventure! Mitchell showed me how to use poles made of cactus ribs to knock the fruit off the top of the tall cacti. It was hard, but I finally got the hang of it. During a water break, I asked Grandma Vavhia what the Tohono O'odham do with the harvested fruit.

"Come sit in the shade" she said, "and I will tell you a story."

"Long ago," she said, "the daughter of a powerful man got caught in a swift dust storm. Her father was furious. He convinced his neighbors to drive away Wind, who had caused the storm. So Wind departed from the O'odham lands. But Wind's blind friend, Rain, followed him. They were both very angry at being treated so rudely.

The plants withered and the animals got thirsty. They asked Hummingbird to find Wind and Rain. Smart Hummingbird tied some of his feathers to a tiny stick, which he carried in his beak. He flew until he saw the feathers blowing in a breeze. He came to a cave where he found Wind and Rain sulking. Hummingbird begged them to return. But Wind was still angry, and Rain could not return by himself. Wind said, 'Tell our relatives they must honor us properly first.' The O'odham prepared delicious saguaro wine called nawait, sang beautiful songs, and told their best stories. After four days, Wind and Rain returned to the O'odham.

"Our ancestors decided to perform the Nawait Ceremony every year to remind us of the importance of Wind and Rain. Every year, we 'pull down the clouds' by harvesting the saguaro fruit. We squeeze out the juice to make wine. The breath we use to tell our stories and sing our songs reminds us of our special relationship with Wind. The nawait we drink reminds us of our sacred relationship with Rain. We need both wind and rain to grow our crops and harvest the fruits of the desert."

17 July

It's hot in the middle of the day here, so we decided to have some night adventures. Last night some friends and I went looking for scorpions! Suddenly, Mitchell flipped on a black light. Aha! There were three scorpions glowing lime green in the black light. They couldn't see the black light and never knew we were there. Standing quietly in the dark, we watched as they scurried around, looking for crickets and other insects to eat. Then, just when we were about to head home, an elf owl swooped down and captured one! What a hoot! On the way home, we stopped to listen to the night noises of the desert. I could hear the calls of the Couch's spadefoots as they headed out for a night swim in the pools created by yesterday's heavy rain. I am amazed that amphibians can survive in the desert!

Stripe Tailed Scorpion
(Vaejovis spinigerus)
LENGTH: 2 INCHES (5 CM)

Appreciating Scorpions:
Tips for the Wary Explorer

Some people are scared of scorpions, but I think they are cool! Just like human moms, mother scorpions drive their kids around until they are old enough to drive themselves. Their eight legs can move over all kinds of desert terrain. After the babies are born, they scramble onto their mother's back, where they ride for the first week or two. But they don't have seatbelts, and if the babies fall off, they could be eaten by bigger animals—or even their mother.

Scorpions are nocturnal. During the day, they spend their time sleeping under rocks, in burrows, or (in the case of the bark scorpion) under the loose bark of trees. The bark scorpion is Arizona's most poisonous scorpion, and on rare occasions has been fatal to humans. The stripe-tailed scorpion is Arizona's most common scorpion. It is much less dangerous than the bark scorpion. Still, it's no fun to get stung, so don't put your hands or feet in hidden places without looking first!

Couch's Spadefoot
(Scaphiopus couchi)
SIZE: 2.25-3.5 INCHES (5.7-8.9 CM)

Spadefoot Survivors:
Outeat, Outdig, Outfast.
by Gretchen Askers, Planet Scout Gazette Staff

If you see an animal that looks like a toad and sounds like a sheep, it's probably a Couch's spadefoot. Spadefoots are in a separate classification from frogs and toads (and especially sheep) because they have a sharp-edged "spade" on the outer edge of each hind foot. Spadefoots use this device to dig backwards, in a circle, into the ground. This is how they survive when it gets too hot or dry for their liking. They can dig as deep as 25" (63.5 cm) below the surface. Down there, the temperature is more constant, and there is more moisture in the soil. The spadefoot can stay underground for up to 10 months because it can eat enough food in one meal to last a whole year! Spadefoots are called out of their burrows by the low-frequency sounds made by thunder and falling rain. Can you imagine the good time they must have in the rain after being underground for that long?

SEASON NOTE:
July is called Rainy Moon because it's the beginning of monsoon season, when violent thunderstorms crash into the desert, filling up empty stream beds (called arroyos) and bringing a second springtime.

GIANT VINEGARONE

BARK SCORPION
Centruroides
exilicauda

GIANT HAIRY
SCORPION
Hadrurus
arizonensis

Mastigo-
roctus
giganteus

PALE
WIND
SCORPION

Eremobates
pallipes

STRIPE
TAILED
SCORPION

Vaejovis
spinigerus

YUCCA GIANT SKIPPER

GRAY METALMARK

Apodemia palmerii

CHARA CHECKER-SPOT

Dymasia chara

ANTILLEAN BLUE

Megathymus yuccae

Hemiargus ceraumus

COMMON SULPHUR

Colias philodice

AMERICAN SNOUT

Lybytheana carinenta

SLEEPY ORANGE

Eurema nicippe

PEARLY MARBLE WING

Euchloe hyantis

American Snout
(Libytheana carinenta)
WINGSPAN: 1 3/8 - 2 INCHES (3.5-5 CM)

21 August
I knew that the desert was famous for its sudden storms, but can you imagine driving into a flash flood of butterflies? They were everywhere! My parents and I were taking a drive and just minding our own business, when we had to pull off the road to avoid a snout butterfly storm. We couldn't see where we were going at all! I watched them swirl past the car window and wondered why there are so many of them here this time of year? And how did they get such a silly name?

Look Out For The Snouts!
By Anne Tenna,
Planet Scout Gazette Staff

There are over 250 species of butterflies in the Sonoran Desert. All of them are colorful and diurnal. (That means they sleep at night and are awake during the day, like us.) As you know I make it my business to study any animal with antennae, or "feelers." This week I set out to find the Sonoran Desert butterfly with the funniest name. After some research I discovered the American snout butterfly, the only snout in the Sonoran. A butterfly with a snout? Apparently it was named for its palpi, the sense organs that stick out of the front of its head. They must have reminded someone of a snout. Look out for the snouts — they can have giant population surges about this time of year!

SEASON NOTE:
August is Short Planting Moon. Late crops are planted. Sometimes there are dramatic thunderstorms in the afternoon, as the monsoon season continues.

10 September

To start off the new school year, my new Planet Scout friends and I took a field trip to see some bighorn sheep in the wilderness near Sabino Canyon. We drove to the edge of a park trail, then hiked a mile up into a rocky pass where the terrain was very steep and hard to climb. I felt like a real explorer! Up on the ledges we got a good look at some bighorns. I counted five sheep. Some were resting and chewing their cuds, while others browsed around for food. By the time we got back to the van, I had finished drinking all the water in my canteen, which made me wonder: Where do the bighorns get their water up here? Our guide explained that if water is not available, they will kick and butt barrel cacti to get at the tender, watery flesh.

Bighorn Sheep
(ovis canadensis)
WEIGHT: MALES 175-275 LBS (79.379-124.74 KG)
FEMALES 250 LBS (68.039 KG)

SEASON NOTE:
September is called Dry Grass Moon because the summer rains stop and the weather gets hot and dry, although the nights are a little cooler. I think it should also be called 'Sticker and Burr Moon' because seeds get stuck all over your shoes and pants. That's how they disperse.

SPIKEDACE   Meda Fulgida

3

SPECKLED
DACE

Rhinichthys
osculus

LONGFIN
DACE

Agosia
chrysogaster

GOLDEN
SHINER

Notemigonus
crysoleucas

MOSQUITO
FISH

Gambusia
affinis

DESERT
PUPFISH

Cyprinodon
macularius

GILA
TOPMINNOW

Poeciliopsis
occidentalis

MEXICAN LONG-NOSED BAT

Leptonycteris nivalis

CALIFORNIA LEAF-NOSED BAT

Macrotus californicus

TOWNSEND'S BIG-EARED BAT

Plecotus townsendii

WESTERN MASTIFF BAT

Eumops perotis

LONG-TONGUED BAT

Choeronycteris mexicana

SANBORN'S LONG-NOSED BAT

Leptonycteris sanborni

WESTERN PIPISTRELLE

Pipistrellus hesperus

YUMA MYOTIS

Myotis yumanensis

Spotted Bat

(Euderma maculatum)

WINGSPAN: 13-14 INCHES
(34-36 CM)

31 October

When Mitchell and I were playing chess this afternoon, I was reminded of the sneaky spotted bat. Why would chess remind me of a bat? Its back actually looks like a chess board, with its bold black and white markings, but that's not the main reason. This bat is a genius at strategy. Every night it checkmates moths with its special-ized echolocation abilities. I bet I could beat it in real chess, though!

P.S. Our Planet Scout Troop is building a special exhibit about desert bats for the Arizona - Sonora Desert Museum. I get to study the tiny spotted bat.

The Chess Game of Bat vs. Moth
By Megan

Bats and moths are locked in a chess game where there have been some amazing moves! Evening bats, those that hunt insects at night, use a special technique called echolocation. The bats send high frequency sounds out into the dark. They use their excellent ears to intercept the echoes that bounce off surrounding objects. By listening, the bats can tell how big an insect is, what shape it is, which direction it's going, and if it's fuzzy or not. Then the bats swoop in to get their meals.

However…

Some moths can hear the bats' echolocation signals. Armed with intelligence of their own, the moths then do one of two things. They either dive into nearby trees or shrubs, which act as cloaking devices, or adopt erratic flight patterns to make the bats think they are not really moths. Some moths go so far as to send a little counterintelli-gence back to the bats. They transmit a signal that sounds like another bat is about to attack them. This causes the real bat to break off the attack, and the moth escapes.

However…

Some bats evolved softer voices and bigger ears to avoid being detected by these clever moths. The spotted bat is currently the reigning superpower in the bat-moth battle—like the winning Queen in chess. It emits sounds that are too low for the moths to hear—even the smart ones. What will moths come up with as their next move?

SEASON NOTE:

Happy Halloween! October is Small Rains Moon, because we have only gotten a bit of rain. Even butterflies are hopping on their brooms this time of year-desert brooms, that is! The desert broom is a shrub that blooms in October. Its flowers attract butterflies. You might see as many as 45 species on one plant!

9 November

What a magical night! My dad and I went out for a walk to see the full moon. It was so bright I could have almost read a book by moonlight! We came over a slight hill and saw a very mysterious thing—a hush of about 20 black-tailed jackrabbits. What were they doing out there in the middle of the night? The dark black sky sparkled with stars. The rabbits seemed to be snacking on the shrubbery. The air was so quiet that I could almost hear them munching in the moonlight. Later, we could hear coyotes talking to each other. Their song was sort of creepy. It made me feel lonely inside, and reminded me of the loon calls where I used to live.

Coyote
(Canis latrans)
WEIGHT: 19.8 - 35.3 LBS.
(9 - 16 KG)

Black-Tailed Jackrabbit
(Lepus californicus)
LENGTH: 18-25 INCHES
(46-64 CM)

CLEVER COYOTES

The coyote is an opportunist. It can live almost anywhere and eat almost anything. The coyote has a habitat range that is more diverse than any other animal in the world, except humans. Their haunting howls followed the pioneers as they traveled into the Wild West centuries ago. Today we have learned to understand their calls a little bit better. At dusk, coyotes sing a "good morning" wake up song before they prepare for the night's hunt. They sing to communicate with their neighbors and family groups, during the full moon, after summer rains, or even just for fun. Coyotes hunt small mammals and insects, and scavenge for dead animals. In the fall and winter months, they include berries, fruit and leaves in their diets. Because their food sources are so unreliable, you will never come across a fat coyote.

Some Hare-Raising Questions
by Peter Abbit, Planet Scout Gazette Staff

Q: How did the black-tailed jackrabbit get its name?
A: Jackrabbits got their names from early settlers, who thought their long ears looked like those of the donkey. These newly discovered animals were thus named "jackass rabbits." This name was later shortened to "jackrabbit".

Q: How are jackrabbits and regular rabbits different?
A: Actually, jackrabbits are part of a group of rodents called hares, which are not the same as rabbits. You could think of a hare as a "rabbit 2.0". Hares are bigger, faster and live longer than rabbits. They also give birth to more fully developed young. Baby hares, called leverets, are born furred, and with their eyes open, whereas baby rabbits, called bunnies, are born blind and naked.

Q: How do black-tailed jackrabbits cope with the desert heat?
A: Like many desert animals, jackrabbits rest during the hottest part of the day. They often seek shelter in the shade of dense vegetation, where they dig shallow depressions.

Q: What do they eat?
A: Black-tailed jackrabbits are plant-eaters. In the summer, they prefer herbaceous plants, when they can find them. In the winter, they eat woody and dry vegetation. Cacti fill most of their water needs, but jackrabbits will drink fresh water if it is available.

SEASON NOTE:
The Pleasant Cold Moon is definitely pleasant - like an early summer day up North - but I wouldn't call it cold. Up in the canyons, leaves are turning orange and yellow and red. Mount Lemmon just got a couple of inches of snow.

GRAY
FOX

DESERT
COTTONTAIL

Sylvilagus
audubonii

KIT FOX
Vulpes macrotis

MEXICAN
GRAY
WOLF

Canis lupus
baileyi

Urocyon
cinereoargenteus

BLACK
TAILED
JACK
RABBIT
Lepus
californicus

ANTELOPE
JACK RABBIT
Lepus alleni

ANDERSON
LYCIUM

RABBIT
BRUSH

Chrysothamnus
nauseosus

SNAKEWEED

Xanthocephalum
sarothrae

Lycium
andersonii

DESERT
SAGE

DESERT HOLLY

Atriplex hymenelytra

DESERT BUCK-WHEAT

Eriogonum fasciculatum

Salvia dorrii

---

28 December

Grandma Vavhia says the creosote bush is one of the longest-lived plants in the world. One stem can live 200 years, but that's not the amazing part. While the two of us were taking a walk, she explained that as the center of the bush begins to decay, it sends new stems out from the edges. The creosote bush spreads out in a circle as it grows. The biggest known ring measures 26 feet wide and may be as old as this "young" desert – perhaps 8,000 years!

Today I saw a cactus wren building a nest in the branches of a cholla (pronounced "CHOY-ah"). Early pioneers believed that the cholla threw its spiny cactus fingers at people walking by. That's because the sections detach easily and can stick to your clothes. The cactus wren's nest looks like a twiggy football. Mitchell said that these birds are compulsive nest-builders. Nests that are not built to house eggs are called roosting nests. I guess that's what this one is.

P.S. I wrote this poem beneath the dripping eaves of our porch during a rainstorm.

Creosote
(*Larrea tridentata*)
HEIGHT: ABOUT 4 FEET (120 CM)

Cactus Wren
(*Campylorhynchus brunneicapillus*)
LENGTH: 7-8.75 INCHES (18-22 CM)

SMELL OF DESERT RAIN ON THE CREOSOTE by Megan

Heat beats hot and dry,

Shining brightly in my eyes.

The meager drops that feed the earth

come slamming down for all they're worth.

They soak the ancient creosote.

Twelve thousand years a ring could stand

Expanding in the desert sand.

When the leaves are wet by rain

I smell the ancient scent again.

SEASON NOTE:
Big cold moon is a much-needed relief from the heat. The days are sunny and cool, and night temperatures only occasionally drop below freezing. It has even rained a few times this month.

10 January
 Today our science teacher invited a docent from the Arizona-Sonora Desert Museum to tell us about some desert species that are endangered or threatened. My favorite was the Gila monster! The Gila monster's skin sparkles a little bit because its scales are round like jewelry beads. Did you know it is one of only two poisonous lizards in the whole world? The other one lives in Mexico.

P.S. Mitchell got me a pet tarantula for my birthday today. I named her Harriet, because she is so hairy. Ha ha! I hope we get along...

Arizona Blond Tarantula
(Aphonopelma chalcodes)
SIZE: 3-4 INCHES (70-100 MM)

Gila Monster
(Heloderma suspectum)
LENGTH: 18-24 INCHES (45.7-61 CM)

Meet the Gila Monster

Gila Habits: During the spring and summer, the Gila monster is active during the day and rests at night. In the winter, it hibernates in a burrow. The females lay their eggs in the fall. There are usually 5 to 12 eggs in a clutch, which overwinter in the nest before hatching next spring.

Monster Measurements: Adult Gila monsters may grow to be anywhere from 18 to 24 inches (45 to 61 cm). They deal with the desert conditions by eating lots of food when it is available, and storing energy in their stumpy tails. Juvenile Gila monsters may eat as much as 50 percent of their body weight in one feeding!

Scarcely Scary: The Gila monster was the first venomous animal protected by law in North America. In 1952 it became illegal to kill, collect or sell them in Arizona. It is one of two known poisonous lizards in the world. (The other is a close relative living in Mexico.) Gila monsters only use their poison defensively. They bite their adversaries, and then chew poison into them. They hiss and back away before they strike. As with all poisonous animals, never handle them unless trained to do so.

SEASON NOTE:
This season is called No More Fat Moon. It got its name because by this time of the year animals have used up their winter fat, and look thinner.

Taming Arizona Blond Tarantulas
By Eric Neds, Planet Scout Gazette Staff

How often should I walk my tarantula?
The good news is that tarantulas don't require walks!

What type of habitat should I prepare for it?
Make a super condo for your tarantula in a small terrarium. Carpet the floor with two to three inches of rocks or soil and keep it damp by squirting it with a spray bottle every three days. Make sure your spider has a nice bedroom. A hollow log might work. Be sure to keep each spider in a separate habitat. Always keep a secure roof on its house. Otherwise, it will escape to find another cozy spot – like under your bed!

What does it eat?
Feed your tarantula crickets, locusts or roaches. Keep a small bowl of fresh water for it to drink. Don't worry about overfeeding; tarantulas only eat what they need.

What temperature does it prefer?
Keep the temperature about 75 F. You can do this by purchasing a heating pad from a pet store. Monitor the tarantula's tank temperature by placing a stick-on thermometer at ground level on the outside of the terrarium. You can buy these thermometers at pet stores too.

Do tarantulas like to play?
Unless you are experienced with tarantulas, we do not recommend handling your fuzzy new spider pet. If disturbed, the Arizona Blond Tarantula's bite can hurt about like a bee sting. Also, it can flick hairs off its back before it resorts to biting. These hairs are very irritating and hard to remove from your skin.

LONGNOSE
LEOPARD
LIZARD

COLLARED LIZARD

crotaphytus
collaris

CHUCKWALLA

saromalus
obesus

Gambelia
wislizenii

WESTERN
BANDED
GECKO

FRINGE-
TOED
LIZARD

uma
notata

Coleonyx
variegatus

REGAL HORNED
LIZARD

Phrynosoma solare

BLACK-CHINNED HUMMINGBIRD

Archilochus alexandri

NORTHERN FLICKER

Colaptes auratus

BROAD-BILLED HUMMINGBIRD

LADDER-BACKED WOOD-PECKER

Picoides scalaris

Cynanthus latirostris

ANNA'S HUMMING BIRD

Calypte anna

RUFOUS HUMMING BIRD

Selasphorus rufus

Costa's Hummingbird

(Calypte costae)
LENGTH: 3.5 INCHES (9 CM)

Host of the Hummingbird House
by Gretchen Askers, Planet Scout Gazette Staff

Today we are here with Abe Podiformes, a docent at the Hummingbird Aviary. Mr. Podiformes, where do hummingbirds live in the wild?
Well, Gretchen, you might be surprised to learn that hummingbirds are only found in the Americas. Scientists have counted over 300 species of them.

Wow! That's amazing! What do they all eat?
Hummingbirds use their slender beaks to sip flower nectar (they eat many small insects as well). They are very fast little birds. They use so much energy that they need to eat almost constantly. Nectar is full of sugar, so it is perfect high-performance hummingbird fuel.

Which hummingbird is best adapted to the dry climate here in Arizona?
Costa's hummingbird is probably the best arid-adapted hummingbird. It does not need to migrate unless there is a food shortage.

I see that this species is making nests during this time of year. Could you tell our readers a little bit about the life history of a Costa's hummingbird?
These birds are highly territorial. Males protect courting zones, females protect nesting areas, and they both guard the feeding territory. The females build nests made of plant down, seeds and mosses. These tiny creations are held together with spider webs and decorated with lichen, leaves, and bark for camouflage. The females lay two eggs that hatch about two weeks later. After a month of being fed by their mothers, the fledglings are all grown up.

SEASON NOTE:
In February, during the Gray Moon, the desert begins to warm up again, with temperatures creeping up into the 80's. Costa's hummingbirds begin to mark their breeding territories around feeders or nectar-laden shrubs, and Gila woodpeckers mark their territories with loud hammering sounds.

---

26 February
Well, I woke up in a strange way today! Rat-tat-ata-tat! I stuck my head out the window, and found that it was a Gila woodpecker hammering a hole in a nearby saguaro. What a racket! Doesn't it hurt the cactus? After school, my parents took Mitchell and me to the Arizona-Sonora Desert Museum, where we visited the Hummingbird Aviary. What a great Friday treat! One hummingbird actually landed on my shoulder and flew away with several red threads in its mouth. Hey, where was she going with my sweatshirt?

Gila Woodpecker

(Melanerpes uropygialis)
LENGTH: 8-10 INCHES (20.3-25.4 CM)

Weekly Beak: Gila Woodpecker
BY BILL HUNTER, PH.D.,
SHARPES INSTITUTE OF BIRD BEAKS
SPECIAL TO THE PLANET SCOUT GAZETTE

At the Sharpes Institute, beaks really pique our interest. This week's featured beak belongs to the Gila woodpecker. This black and white striped bird uses its beak to gather food, make nesting holes, and even to mark territory.

This bird eats saguaro fruit, insects, mistletoe berries, and nectar. It will even steal dog food! As an eating utensil, the Gila woodpecker's beak is the bird equivalent of a spork.

Gila woodpeckers use their powerful beaks to hammer nest holes into the mighty saguaro. They rarely dig deep enough to hurt the cacti, though. You'd think that this hammering would make the woodpeckers dizzy. Imagine trying to dig a hole in a cactus with your nose! Fortunately, woodpeckers have a surprising adaptation that protects their bird brains. They have long tongues, which they roll up inside their heads to absorb the shock.

Most birds mark their territories with song. Instead, Gila woodpeckers tap loudly on metal objects such as rain gutters to mark their territories. The noise can be bothersome if you are trying to sleep!

15 March

It seems like the sand itself is blooming, the flowers appeared so quickly! Dancing down the dunes, or rocketing up from rocky outcrops, are dramatic drifts of desert flowers. (I am getting good at alliteration!) I danced too, spinning around and around until I fell down laughing in a bed of flowers! I imagined the flowers creeping out from under rocks, and climbing up to the top of the legume trees. There's the mariposa lily, the desert anemone, and my favorite, the desert lily, which seems to shoot up to the sky out of nowhere. Grandma Vavhia says the Desert Lily can occasionally grow to be six feet tall in a really wet year. How does it survive during dry years?

Desert Lily
(Hesperocallis undulata)
FLOWERS: 2 INCHES LONG (5 CM)
WHITE, BLOOM IN LATE SPRING

Habitat: Loose, sandy soils, along desert roads

Range: Eastern Mojave Desert south through Arizona to Northern Sonora, Mexico and central Baja California

Leaves: The desert lily sprouts a cluster of long, wavy leaves. They are blue-green white edges, and can be 8 to 20 inches (20-50 cm) long.

Size: The flower spike can be from 1 to 6 feet (30 cm – 2m) tall, depending on the abundance of water in a season. During dry periods, the leaves die back, and the plant becomes invisible from above ground.

Arid Adaptation: The desert lily behaves like a spring annual. It seems to sprout from nowhere every spring. In fact, it grows from a bulb that is buried 2 feet (0.6 m) underground. This protects it from hungry animals, scorching surface temperatures, and heavy summer rains that might rot the bulb.

Planet Scout Flower Finder 72

ARIZONA BLUE-EYES

DESERT FOUR O'CLOCK

Mirabilis multiflora

INDIAN BLANKET

Gaillardia pulchella

GHOST FLOWER

Mohavea confertiflora

ANGEL TRUMPETS

Acleisanthes longiflora

DESERT
GLOBEMALLOW

Sphaeralcea ambigua

DESERT
FIVE
SPOT

Malvastrum rotundifolium

PURPLE MAT
Nama demissum

ulus
icus

Zephyranthes longifolia

WOOLY
DAISY

Eriophyllum wallacei

DEVIL'S
CLAW

Proboscidea altheaefolia

YELLOW
MUD TURTLE
kinosternon
flavescens

BARN OWL
tyto
alba

WESTERN
BOX
TURTLE

terrapene
ornata

WESTERN
SCREECH-
OWL

otus
kennicottii

BURROWING
OWL

athene
cunicularia

SPINY
SOFTSHELL

trionyx
spiniferus

SONORAN MUD TURTLE

kinosternon
sonoriense

FERRUGINOUS
PYGMY-OWL
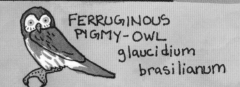
glaucidium
brasilianum

Elf Owl
(Micrathene whitney)
Size: 5-6 INCHES
(13-15 CM)

Mitchell gave me a copy of this poem he wrote for his science project. Maybe I should write more poetry myself...

Guess Hoo?
by Mitchell

Kew kew kew.
Who's chirping at you
From a hole in the giant Saguaro?

It's the size of your fist,
And I really insist,
It's the smallest owl in the world.

You'll be happy to know
It eats bugs as it goes
To follow them south for the winter.

If you still couldn't say
Who's the bird of the day,
I will end the suspense-
Please be its defense!
It's the Endangered and tiny Elf Owl! .

21 March
I was on my way to observe the Desert Tortoise at the Desert Museum for a science report, when I got the feeling I was being watched. I noticed a tiny bird staring at me from a hole in a Saguaro. I asked a docent, who told me it was an Elf Owl. That's the same owl that captured the scorpion! Do they make those nest holes themselves? That's the third animal I have seen that lives in a Saguaro! I snapped some photos of the Desert Tortoise, and watched him chomp a bright red flower for lunch. Hey, I wonder if a turtle is the same thing as a tortoise. That would be a good thing to put in my report.

Desert Tortoise
(Gopherus agassizi)
Size: 9.25-14.5 INCHES
(23.5-36.8 CM)

Tortoise Time:
The Desert Tortoise is a very long-lived species. The shell of a baby tortoise does not completely harden until it turns 5 years old. They reach maturity at 15. Their life span is about 35 to 40 years, but it is said that they can get to be 100 years old.

Tortoise and Turtle:
A Tortoise is a type of turtle. You can tell a tortoise from a turtle by the shape of its feet. Tortoises have short, stumpy feet that are better for walking on land and digging burrows. Turtles spend more time in the water, so their feet are better for swimming.

SEASON NOTE:
Melted snow rushes down the mountains to water the desert during the Green Moon. Dry creek beds fill up and the desert is briefly refreshed. The annuals begin to bloom, then the perennials and shrubs.

17 April
Until today, there was one bird that I hadn't seen yet. It was the hero of my favorite cartoon. The roadrunner! (Beep beep!) Today I spotted one running with a lizard in its beak. It didn't look anything like the cartoon, but it sure was speedy. How fast can it run?

P.S.: I saw a hedgehog yesterday, and today on a trip further west in the desert I saw a beavertail. It's not what you're thinking, though – they were covered with bright pink and purple flowers (respectively)! These aren't animals, but funny names for cacti.

Greater Roadrunner
(Geococcyx californianus)
Size: 20=24 inches (51-61 cm)

ASSIGNMENT:
Fill in the blanks with phrases describing the roadrunner.
Think about what it eats, how it moves, and where it lives.

R UNNING RAPIDLY

O VER ROADS AND DESERT

A CHIEVING SPEEDS OF 15 MILES PER HOUR

D INING ON INSECTS, SCORPIONS, LIZARDS AND SNAKES

R APID RELATIVES OF THE CUCKOO

U TTERING A SONG OF LOW COOING NOTES

N OTICE THE POSSIBLY PERMANENT PAIR BOND

N ESTING IN CHOLLA, MESQUITE, OR PALO VERDE

E VEN DEFENDING TERRITORY ALL YEAR

R UDDER-LIKE TAILS

S ELDOM FLIES

CACTUS CARDS: Strawberry Hedgehog

Common English Name: STRAWBERRY HEDGEHOG
Common Spanish Name: CACTO FRESA
Scientific Name: ECHINOCEREUS ENGELMANNII
Size: CLUSTERS OF UP TO 60 1 FOOT (30 CM) TALL STEMS
Flowers: DEEP RED-PURPLE TO LAVENDER, APRIL
Fruit: EDIBLE, WITH JUICY RED PULP
Spines: SPROUTING FROM EACH AREOLE ARE 2 - 6 LONG, CENTRAL SPINES, WITH 6 - 14 SHORTER RADIAL SPINES.

CACTUS CARDS: Beavertail

Common English Name: BEAVERTAIL CACTUS
Common Spanish Name: NOPAL
Scientific Name: OPUNTIA BASILARIS
Size: 6 FOOT (1.8 M) DIAMETER CLUMPS, 1 FOOT (30 CM) TALL
Flowers: 2-3", INCANDESCENT PINK, LATE FEBRUARY TO MAY
Fruit: 1.25" (3.1 CM) LONG, EGG-SHAPED, GREYISH BROWN, INEDIBLE, DRY, WITH VERY LARGE SEEDS
Spines: GLOCHIDS (TINY IRRITATING HAIR-LIKE SPINES). THESE PRICKLING PESTS LOOK SOFT, BUT DO NOT TOUCH THEM.

SEASON NOTE:
April is called Yellow Moon because this is when the legume trees such as the palo verdes bloom. Their flowers range in color from light cream to bright yellow.

FISH HOOK
BARREL

ENGELMANN
PRICKLY
PEAR

Opuntia
engelmannii

CLARET
CUP
CACTUS

Echinocereus
triglochidiatus

RAINBOW
CACTUS

Ferocactus
wislizeni

Echino-
cereus
pectinatus

COVILLE
BARREL
Ferocactus
emoryi

MANY-HEADED
BARREL

Echinocactus
polycephalus

PHAINOPEPLA   Phainopepla nitens

GAMBEL'S QUAIL   Callipepla gambelii

JAVELINA

Tayassu tajaou

SPECKLED RATTLE SNAKE   Crotalus mitchelli

COLLARED LIZARD

Crotaphytus collaris

'SMORES

BOOK

7 May
 Spring Break gave me the opportunity to explore the desert up close. My family planned a three-night camping trip in the Rincon Mountain District of the Saguaro National Park. I was thrilled to show my parents how much I had learned since we moved to the desert!

Our Camping Adventures:

Day One-
• hike to Douglas Springs Campground (6 miles, elevation 4,500 feet)
• stop for rattlesnake crossing and water breaks
• set up camp and build fire, make dinner
• hop into sleeping bag + finish exciting chapter book I'm reading!

Day Two-
• short bird walk at sunrise before breakfast
(saw a bunch of Gambel's Quails and a couple of Phainopeplas!)
• hike into cactus forest with pack lunch
(saw a pack of javelinas snacking on some prickly pear pads!)
• back to camp, dinner and campfire with 'smores
• look at stars (you can see so many of them away from the city!)

Day Three-
• Pack up camp, hike out of park
(I pointed out a well-hidden Collared Lizard just off the trail!)

After searching all year, I finally found my new special spot! Here in the arms of a big tree near our campsite, I can see miles of desert below me. Now the desert really feels like home. Soon school will be out and we can start harvesting saguaro fruit again. Then a new desert year will begin!

SEASON NOTE:
In May the Sonoran Desert really starts to heat up. We had our first day over 100 F (38 C) since last summer. While cactus giants such as the Saguaro are opening their nocturnal blossoms, animals on the ground begin to run out of resources. This season is called Painful Moon because the traditional O'odham are getting hungry again.

So what can I do?

1. Read a Book!

A Desert Scrapbook: Dawn to Dusk in the Sonoran Desert
Written and Illustrated by Virginia Wright-Frierson
Simon & Schuster Books For Young Readers, New York, NY
32 Pages, © 1996

Alejandro's Gift
Written by Richard E. Albert
Illustrated by Sylvia Long
Chronicle Books, San Francisco, CA
32 Pages, © 1996

Sing Down the Rain
Written by Judi Moreillon
Illustrated by Michael Chiago
Kiva Publishing,, Walnut, CA
32 Pages, © 1997

Cactus Poems
Poems by Frank Asch
Photos and Notes by Ted Levin
Gulliver Green, San Diego, CA
48 Pages, © 1998

Deserts: A National Audubon Society Nature Guide
Alfred A. Knopf, New York, NY
640 Pages, © 1998

Arizona-Sonora Desert Museum Book of Answers
By David Wentworth Lazaroff
Arizona-Sonora Desert Museum Press, Tucson, AZ
192 Pages, © 1998

A Natural History of the Sonoran Desert
Edited by Steven J. Phillips & Patricia Wentworth Comus
Arizona-Sonora Desert Museum Press, Tucson, AZ
628 Pages, © 2000

2. Get on the Web!

The Arizona-Sonora Desert Museum
www.desertmuseum.org

Desert USA
A learning and travel site covering all 4 American deserts.
www.desertusa.com

The Owl Pages-Information about Owls
List of links to Owl Websites, and a cool "live nest-cam" of a
Great Horned Owl's nest.
www.owlpages.com

One World Journeys
www.oneworldjourneys.com/sonoran

3. Join Planet Scouts!

You can become a Planet Scout too!

Drop by the Planet Scout Clubhouse and make some new friends, learn how to keep your own journal and even read about the adventures of other Planet Scouts.

www.planetscouts.org

DEAR PLANET SCOUTS—

Hi! My name is Kristin. I am the author and illustrator of this book. Probably each of us has a special dream that we hope to fulfill someday. My favorite thing to do is draw pictures and write stories. I don't think anyone is too young to start making their dreams happen. Just show up and do your best! This is my fifth book. I hope you like it!

♥, Kristin

kristin@planetscouts.org

A Note About Making This Book:
The illustrations in Saguaro Moon are watercolor paintings. The story was written on a tangerine Apple iBook laptop. Quite a few fonts were used in this book: the Cholla and Base 12 font families by Emigre, Berkeley Oldstyle and Courier from Adobe, Tarragon from Letraset and MeganMots, a new font created specifically for this book by Gabriel Serafini using Megan Overton's handwriting. The artwork was scanned, along with all the notes, articles and backgrounds and assembled in Adobe Photoshop on an Apple Macintosh G-4.

Thank You!

I would like to thank the following people for their support and encouragement during the creation of Saguaro Moon:

- Mark Dimmitt
Director of Natural History,
Arizona-Sonora Desert Museum

- Gabriel Serafini
my husband

- Katie Pratt
my sister

- Kathy Pratt
my mother

- Ken Pratt
my father

- Kevin Pratt
my brother

- Glenn Hovemann, text editor,
Muffy Weaver, art editor
and the rest of the Dawn Publications Staff

- Megan Overton
my cousin who helped me out by lending her
handwriting to the character in the story.

As a teenager, Kristin's pen and paintbrush were passionate about nature. As a result, she had three best-selling books published while she was still a teen. Whether she knew it or not at the time, keeping a journal—no matter how informal—is an extremely effective educational technique. Now, as an environmental educator in her own right, Kristin Joy Pratt-Serafini is modeling how to keep a fun-filled nature journal and encouraging young people to discover a variety of habitats. Her first "model" nature journal was Salamander Rain: A Lake and Pond Journal. With this book Kristin explores a dryer, hotter habitat.

At 14, Kristin wrote and illustrated A Walk in the Rainforest. At age 16 she followed with the popular A Swim through the Sea. At 18 she did A Fly in the Sky. In college, Kristin studied art. Now married with the last name of Pratt-Serafini, Kristin is a self-assured writer and artist, very environmentally aware, whose work is still young and fun.

A Note to Teachers: to facilitate using picture books in the classroom, Dawn Publications offers Teacher's Guides for each of the books in Kristin's Walk/Swim/Fly trilogy (and those of several other authors as well). These 48-page guides offer lesson plans for grades 3 to 6 that are distinctive in that they integrate core science and language arts curricula with character education lessons.

Some Related Books from Dawn Publications

Salmon Stream by Carol Reed-Jones. By the author of The Tree in the Ancient Forest, this book is lively and rhythmic, rich in imagery, yet well founded in the scientific cycle of salmon. It engages children in a positive way, and shows how they can help make sure salmon will always be with us.

The Tree in the Ancient Forest by Carol Reed-Jones. A single old tree is home for owls and voles, squirrels and martens, and many other creatures of the deep woods. Through cumulative verse this book tells the tale of their community—a wonderful introduction to the interdependence of nature.

My Favorite Tree, by Diane Iverson. Each major family of indigenous North American trees is illustrated with a child engaged with it in some way, along with a map and a variety of information about its major features, its role in history, its wild animal companions, and other interesting facts.

This Is the Sea that Feeds Us, by Robert F. Baldwin. In simple cumulative verse, beginning with tiny plankton, "floating free," this book explores the oceans' fabulous food chain that reaches all the way to whales and humans in an intricate web.

A Drop Around the World, by Barbara McKinney. Follow a drop a water in its natural voyage around the world, in clouds, as ice and snow, underground, in the sea, piped from a reservoir, in plants and even in an animal. Science of the water cycle and poetic verse come together in this book. Teacher's Guide available.

Dawn Publications
P.O. Box 2010
Nevada City, CA 95959
800-545-7475
nature@dawnpub.com
www.dawnpub.com

Dawn Publications is dedicated to inspiring in children a deeper understanding and appreciation for all life on Earth. To view our full list of titles or to order, please visit our web site at www.dawnpub.com, or call 800-545-7475.

Library of Congress Cataloging-in-Publication Data
Pratt-Serafini, Kristin Joy.
Saguaro moon : a desert journal / written and illustrated by
Kristin Joy Pratt-Serafini.— 1st ed.
   p. cm. — (A Sharing nature with children book)
   Summary: When her family moves to the Sonoran Desert
   in Arizona, Megan keeps a nature journal in
   which she describes the desert, the changes that occur
   throughout the seasons, and how these affect the plant
   and animal inhabitants.
   ISBN 1-58469-036-4 (pbk.) — ISBN 1-58469-037-2
   (hardcover)
   1. Natural history-Sonoran Desert-Juvenile literature. 2.
   Sonoran Desert-Juvenile literature. [1. Sonoran Desert. 2.
   Natural history-Sonoran Desert. 3. Desert ecology. 4.
   Ecology.] I. Title. II. Series.
   QH104.5.S58 P725 2002
   508.3154'09791—dc21
                                    2002003969

Printed in Korea

10 9 8 7 6 5 4 3 2 1
First Edition
Design by Kristin Joy Pratt-Serafini
Additional design and computer production
by Andrea Miles

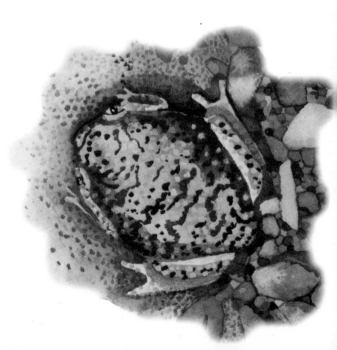